DATE DUE			
T. Smith			
T. Agro	9/8		
SE 1 7 '06	3K		
JA 2 6 '07	1 H		
FEB 0 4 2015	2 m		
MAY 1 2 2015	1 H		

GUMDROP BOOKS - Bethany, Missouri

Snow and Ice

See for Yourself

Snow and Ice

Kay Davies and Wendy Oldfield
Photographs by Robert Pickett

RSVP
RAINTREE
STECK-VAUGHN
PUBLISHERS
The Steck-Vaughn Company

Austin, Texas

Published by Raintree Steck-Vaughn Publishers, an imprint of Steck-Vaughn Company

Editor: Kathy DeVico
Project Manager: Julie Klaus
Electronic Production: Scott Melcer

All photographs by Robert Pickett except pp. 3, 21
© Marcia W. Griffen/Animals Animals

Library of Congress Cataloging-in-Publication Data
Davies, Kay.
    Snow and ice / Kay Davies and Wendy Oldfield; photographs by Robert Pickett.
       p.   cm. — (See for yourself)
    Includes index.
    ISBN 0-8172-4042-X
    1. Snow—Juvenile literature   2. Ice—Juvenile literature.
  [1. Snow—Experiments.   2. Ice—Experiments.   3. Experiments.]
  I. Oldfield, Wendy.   II. Pickett, Robert, ill.   III. Title.
  IV. Series.
  QC926.37.D38   1996
  551.57'84—dc20   95-6007
                                         CIP
                                         AC

Contents

Winter Snow

In the winter, it can get very cold.
Sometimes everything outside
gets covered in snow.

If you go outside, you need to dress
in lots of clothes to stay warm. Your
hands and feet can get very cold.

Look at how the girl in the picture
is dressed.

Which clothes do we wear on
cold days?

Which things keep our hands
and feet warm?

What would you wear on
a snowy day?

Freezing Water

Sometimes the air gets cold. This can make the water cold. If water gets very cold, it will turn into ice. When this happens, we say that water freezes.

Look at the picture of icicles.
As water drips from windows
and roofs, it freezes.
This makes icicles.

You can make ice yourself.
Pour some water
into a plastic bottle.
Put it in the freezer,
and wait.

What has happened
to it by the next day?

Snowflakes

Have you ever seen snow? Snowflakes are
frozen drops of water. They fall from clouds.
On the ground, snowflakes stick together.
They make a blanket of snow.

Look at the single snowflake.
How many points does it have?

How about the ones
in this picture?

All snowflakes have the same
number of points. But each
snowflake is different from
every other snowflake.

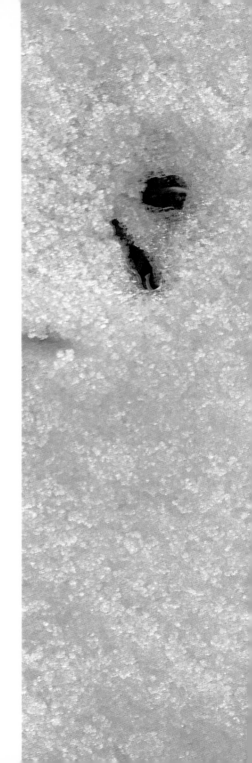

Soft Snow

Snow is soft. When you walk in it, the snow under your feet gets pressed into shapes. These shapes are called footprints.

Animal footprints are called tracks. Can you guess what kind of animal made the tracks in the big picture?

Tracks can tell a story. Look at this picture.

The bird that made these marks was trying to land in the snow. Can you see the shape of its wings in the snow? Can you tell where its feet were?

Slippery Ice

Have you ever seen a frozen pond?
The ice on top is smooth and slippery.

Birds find it difficult to walk on ice.
Look at the seagull in the big picture.
It is trying to keep its balance by
flapping its wings.

You can wear special
boots to walk on icy
surfaces. They have
ridges on the bottom
to help you grip.
(But remember never
to walk on a frozen
pond. It isn't safe.)

Which pair of boots
would you wear to
go walking in snow
and ice?

Why Ice Floats

Try putting some ice cubes
in a glass of water.
What happens?

Ice is lighter than water,
so it floats.

Icebergs are huge lumps of
ice that float in cold oceans.
They can be dangerous to
ships. Sailors cannot see the
part of the iceberg that is
under the water. So they
might crash into it.

When a pond has frozen,
why is the ice always at
the top?

Plants in the Winter

Some trees lose their leaves before winter.

The trees in the big picture look dead, but they are just resting. When spring comes, they will grow new leaves.

Many plants do die in the winter. But some plants live as seeds or bulbs. They rest under the ground in the cold soil. When the warm weather comes, they will start to grow.

Put some mustard seeds into two pots of soil. Put one of them in the refrigerator. Put the other in a sunny window, and water it.

What happens to each pot?

How Animals Stay Warm

If you were cold, you might put on another shirt. But animals can't do that. Look at the picture of the cardinal in the snow. Can you see that its feathers are all fluffed out? Warm air gets trapped between its feathers and keeps the cardinal warm.

Before the winter, some furry animals spend more time in their nests and burrows. This is where they stay warmest.

Can you see this wren at its nest?

Do you think the wren would be warm inside its nest?

Finding Food

It is hard for birds to find food in the winter.
The blackbird in the big picture is looking for food
in the snow. Many of the insects and plants that
birds eat die in cold weather. If ponds freeze, birds
have to find other fresh water to drink.

Some birds fly away to
warmer countries for the
winter. They go where
there are lots of insects
for them to eat.

You can give birds water
and food. Mix nuts and
seeds with some animal
fat. Then put the mixture
into something shaped
like this bell. Let the
mixture harden.

Hang it up outside with a piece of string.
How many different birds come to eat?

Watch Ice Melt

When the weather gets warmer, snow and ice turn back into water, or melt.

You can make ice, and then watch it melt. Find a rubber glove. Pour some water into it. Put a rubber band tightly around the top of it.

Place it in the freezer. Wait until the next morning. Take it out, and remove the rubber band. Ask an adult to help you peel the glove off. Look what you have made!

Will it melt faster in a warm place? Or a cold place? See if you can find out.

Sprouting Seeds

Look at the river in this picture.
The river banks are covered in
snow. As the snow melts, more
water will run into the river.
If the snow melts very quickly,
a lot of water will run into
the river at once. The river
may overflow.

Look at the big picture. It is a
seedling in soil. When snow
and ice melt, some of the water
soaks into the soil. This makes the
soil wet enough for seeds to grow in.

What else happens in the spring when snow
and ice melt?

More Things to Do

1. Make paper snowflakes.
Carefully fold a paper circle six times. Cut patterns into your folded paper with scissors. Unfold it, and you will have a six-pointed snowflake! How many different snowflakes can you make?

2. Make prints in modeling clay.
You can make prints in modeling clay. It is soft like snow, and will show the shape of an object pressed into it. Try squashing keys, buttons, or string into the clay. Coins and shells will make good prints, too.

3. Sayings about snow and ice.
Talk about the meanings of these sayings:

To skate on thin ice.
A heart as cold as ice.
To break the ice.
The tip of the iceberg.

To be snowed under.
As white as snow.
To be snowed in.

Do you know what this is?

You may see them most often in the winter.

Index

This index will help you find some
of the important words in this book.

Notes for Parents and Teachers

As you share this book with children, these notes will help you explain the scientific concepts behind the different activities, and suggest other activities you might like to try with them.

Pages 8–9
When water freezes, it expands. Ice takes up more space than water. A good activity to show this is: put a plastic bottle completely full of water in the freezer with a loose foil cap over the top of the bottle. When the water in the bottle has frozen, the ice will push the cap away from the neck of the bottle.

Pages 10–11
Snowflakes are made up of six-sided ice crystals. The shapes and sizes of snowflakes depend on the height and temperature at which they are formed, and the amount of moisture in the cloud. Looking at pictures of paper snowflakes provides an opportunity to explore the idea of rotational symmetry. When the children make paper snowflakes, you can show them that their snowflake will look the same, whichever one of its points is at the top.

Pages 12–13
When snow is pressed together, the crystals are crushed to form compacted ice. This happens when you squash snow together to make a snowball. Glaciers in the mountains form in the same way: layers of snow are pressed together to make rivers of ice.

Note: The tracks in the picture on page 13 were made by a rook, a bird very similar to a crow.

Pages 14–15
The reason it is easy to slip on a smooth surface, like ice, is that very little friction is created when you walk. Friction occurs when two surfaces rub against each other. Tiny bumps and holes in the surfaces cling together and make things "stick" to each other. Without friction, we would slip and fall over every time we tried to walk. The ridges on shoes or boots help create enough friction for us to grip when walking on smooth surfaces.

Pages 18–19
During the spring and summer, trees take in water through their roots and lose it through their leaves. A medium-size tree can move 26 gallons of water in a day. In the winter, trees cannot draw enough water from the soil to replace the water lost through their leaves. So in the fall, broad-leaved trees shed their leaves before the winter weather arrives.

Pages 20–21
Birds fluff up their feathers in order to trap layers of air between them. Similarly, air is trapped between layers of clothing and helps keep heat in. Several layers of thin clothing will insulate more effectively than one thick piece of clothing. Children can investigate this for themselves by wearing different numbers of layers of clothing.